D0116290

Cheryl Jakab

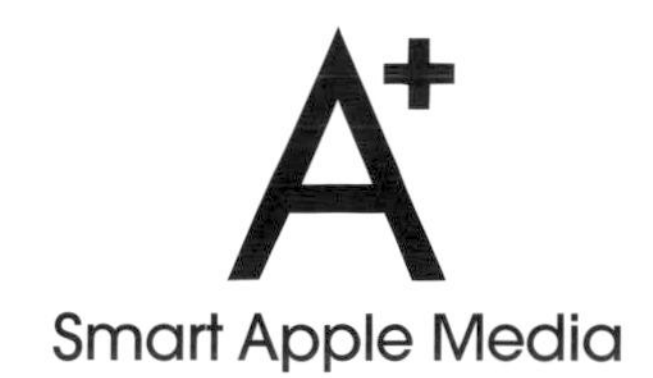

This edition first published in 2010 in the United States of America by Smart Apple Media.

Smart Apple Media
P.O. Box 3263
Mankato, MN 56002

First published in 2009 by
MACMILLAN EDUCATION AUSTRALIA PTY LTD
15–19 Claremont Street, South Yarra, Australia 3141

Visit our Web site at www.macmillan.com.au or go directly to www.macmillanlibrary.com.au

Associated companies and representatives throughout the world.

Library of Congress Cataloging-in-Publication Data

Jakab, Cheryl.
Water supply / Cheryl Jakab.
p. cm. – (Global issues)
Includes index.
ISBN 978-1-59920-456-7 (hardcover)
1. Water-supply–Juvenile literature. 2. Water conservation–Juvenile literature. 3. Water rights–Juvenile literature. I. Title.
TD348.J35 2010
363.6'1–dc22
2009002025

Edited by Julia Carlomagno
Text and cover design by Cristina Neri, Canary Graphic Design
Page layout by Christine Deering and Domenic Lauricella
Photo research by Jes Senbergs

Printed in the United States

Acknowledgments
The author and the publisher are grateful to the following for permission to reproduce copyright material:

Front cover photograph: Aerial view of aqueduct, California, © iofoto/Istockphoto

Photos courtesy of: Rob Cruse, **29**; © Diademimag/Dreamstime.com, **16**; © Madelaide/Dreamstime.com, **5**; © Aschwin Prein/Dreamstime.com, **12**; © Wormold/Dreamstime.com, **19**; DW Stock Picture Library, **10**, **23**, **26**; Gregg Eckhardt, **22**; Envisat (ESA), **11**; AFP/Getty Images, **27**; Landwater WA, **7** (bottom), **9**; Photolibrary/ © Ace Stock Limited/Alamy, **8**; Photolibrary/ © Afripics.com/Alamy, **20**; Photolibrary/ © America/Alamy, **7** (middle), **24**; Photolibrary/ © Robert Harding Picture Library Ltd/Alamy, **7** (top), **17**; Photolibrary/ © Bill Heinsohn/Alamy, **6** (right), **21**; Photolibrary/ © Image of Africa Photobank/Alamy, **6** (left), **13**; Photolibrary/ © Per Karlsson/BKWine.com/Alamy, **18**; Photolibrary/ © Jerry & Marcy Monkman/EcoPhotography.com/Alamy, **15**; Photolibrary/ © Helene Rogers/Alamy, **25**; Photolibrary/ © Kevin Schafer/Alamy, **14**.

While every care has been taken to trace and acknowledge copyright, the publisher tenders their apologies for any accidental infringement where copyright has proved untraceable. Where the attempt has been unsuccessful, the publisher welcomes information that would redress the situation.

Please note
At the time of printing, the Internet addresses appearing in this book were correct. Owing to the dynamic nature of the Internet, however, we cannot guarantee that all these addresses will remain correct.

Contents

Glossary Words

When a word is printed in **bold**, you can look up its meaning in the Glossary on page 31.

Facing Global Issues

Hi there! This is Earth speaking. Will you spare a moment to listen to me? I have some very important things to discuss.

We must face up to some urgent environmental problems! All living things depend on my environment, but the way you humans are living at the moment, I will not be able to keep looking after you.

The issues I am worried about are:

- the effects of **global warming**
- the health of natural environments
- the use of **nonrenewable** energy supplies
- the environmental impact of unsustainable cities
- the build-up of toxic waste in the environment
- a reliable water supply for all

My global challenge to you is to find a sustainable way of living. Read on to find out what people around the world are doing to try to help.

Fast Fact

Sustainable development is a form of growth that lets us meet our present needs while leaving resources for future generations to meet their needs too.

What's the Issue? Water Supply Crisis

Freshwater supplies for people on Earth are running low. The demand for water is increasing as populations grow. **Climate change** is altering rainfall patterns and adding to the problem.

Rising Water Use

The demand for water in homes, farms, and industries is rising even faster than the population is growing. While many people are using more water, about one billion people still live without any safe drinking water or **sanitation**. Technologies to increase water supplies in these areas can be costly and are often only available for use in **developed countries**.

Fast Fact

About one-third of Earth's population lives in countries with little or no supplies of clean, freshwater.

Available Freshwater

Most of Earth's freshwater cannot be harvested. Much of it is permanently frozen in icecaps and glaciers. The Amazon River Basin in South America has large amounts of freshwater that are too remote to access.

Climate change is altering rainfall patterns and this is also affecting freshwater supplies. Rain is falling at different times and in different places than in the past. Heavy seasonal rains can cause flooding, rather than add to water supplies. River flows and water stores are declining in many areas. Pollution is also **contaminating** water supplies in rivers and lakes.

Dry, cracked riverbeds such as this can be found in areas where little or no rain falls.

Water Supply Issues

The most urgent water supply issues around the globe include:

- decreasing water supplies due to climate change (see issue 1)
- dams and pipelines causing environmental damage (see issue 2)
- **irrigation** water being wasted (see issue 3)
- overuse of **ground water** (see issue 4)
- unequal access to freshwater (see issue 5)

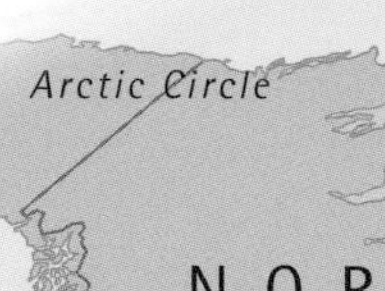

ISSUE 4

United States
Large amounts of ground water from the Ogallala **aquifer** is being used to water crops. See pages 20–23.

ISSUE 2

Lake Victoria, Uganda
The building of the Nalubaale Power Station and dam have caused water levels to drop. See pages 12–15.

Around the Globe

EUROPE
ASIA
AFRICA
AUSTRALIA
Central Asia
Dubai
Lake Victoria
Perth
Tropic of Capricorn

ISSUE 3

Central Asia
The Aral Sea is drying up as too much water is taken for irrigation. See pages 16–19.

ISSUE 5

Dubai
Water is being wasted on swimming pools and golf courses, while some people have little access to freshwater. See pages 24–27.

ISSUE 1

Perth, Australia
Rainfall patterns are changing due to climate change. See pages 8–11.

Fast Fact
The United Nations named 2003 as the International Year of Freshwater.

ISSUE 1

Decreasing Water Supplies

Water supplies are decreasing due to changing rainfall patterns and large amounts of water being harvested. Water-supply systems that have worked well in the past are now failing to meet people's needs.

Changing Rainfall Patterns

Rainfall patterns are changing due to global warming. Some areas on Earth are getting more rain, while others are getting less. The rainy seasons are also changing in many places. Summer storms have replaced heavy winter rains in many **temperate zones**. Summer storms create less **runoff** than winter rains, so less freshwater is available in these areas.

Harvesting Freshwater

Harvesting freshwater is reducing water supplies and damaging local ecosystems. Freshwater is harvested from:

- water **catchments**
- rivers and wetlands
- aquifers and other underground water sources

Most harvested water is used by agriculture and industry. About 10 percent of harvested water is used in homes.

Water supplies are decreasing in many water catchments around the world.

Perth has built a desalination plant to combat the city's declining natural supplies of freshwater.

CASE STUDY

Perth's Decreasing Water Supply

Freshwater supplies are decreasing in the city of Perth, Australia. Rainfall is declining and the freshwater aquifer that lies under the city is slowly drying up.

Less Rainfall

Perth's average rainfall has dropped by at least 15 percent since the 1970s. Records of Perth's rainfall patterns, dating back to 1975, show that more rain now falls as light showers, which creates less runoff. This has led to a dramatic decrease in the flow of water into Perth's dams and aquifers.

Fast Fact

The **aqueduct** system in Ancient Rome supplied water at approximately the same rate as Perth's current water-supply system.

Aquifers Drying Up

In the past, most of Perth's freshwater was pumped from a 40,000-year-old aquifer that lies beneath the city. Today, Perth's population is growing rapidly due to the booming mining industry. As the population grows, this aquifer is being pumped dry.

Today, about one-quarter of Perth's water supply comes from 11 dams and reservoirs. These are connected by a series of pipes that stretch over 47 miles (75 km) through the Darling Ranges. A **desalination plant** has also been added to Perth's water-supply system. It now provides nearly one-quarter of Perth's freshwater.

ISSUE 1

Toward a Sustainable Future: Sustainable Water Use

Water supplies need to be used sustainably. If people know how much freshwater is available on Earth, they can develop systems to manage this water.

Fast Fact
Earth's sea levels are rising due to global warming. If they continue to rise, freshwater supplies in coastal areas may become contaminated by seawater.

Monitoring Water Supplies

People need to be aware of how freshwater supplies in their area are changing so that they can plan for the future. Climate change is changing river flows and rainfall patterns, which affect the amount of ground water in aquifers. If people monitor how water supplies are changing due to climate change, they can more accurately predict how much water will be available in the future.

Managing Water Supplies

Good management of water supplies involves:

- assessing water reserves regularly to note changes in water levels
- limiting the amount of water harvested for agriculture and industry
- regulating water use in industry and homes
- keeping river ecosystems healthy by introducing **environmental flows**
- controlling sources of water pollution to keep freshwater supplies **potable**

Some governments are also raising water prices to reduce water **consumption**.

If freshwater is used sustainably, there will be enough to keep river ecosystems healthy and supply people with water.

Satellite maps of water resources in Africa (shown on this map in green) help Zambia to plan its water use.

CASE STUDY

Monitoring Zambia's Water Supplies

The European Space Agency (ESA) is providing information on water resources in Zambia, Africa, to help improve its water management.

Satellite Mapping

Satellite maps of Zambia's water resources are being created as part of the ESA's Integrated Water Resource Management project. Satellites are mapping surface and underground water reserves. The satellite maps provide information which will allow Zambia to plan for sustainable water use. The project is one of 15 across Africa looking at stages of the water cycle.

Fast Fact

Ninety-five percent of untreated waste in many **developing countries** flows into rivers. This can cause diseases when people use the river water for washing and drinking.

Planning for the Population

Zambia has a large **urban** population. Today, about one-third of the country's 11 million people live in urban areas. Urbanization has occurred faster than services have developed. The ESA's satellite images will help Zambia to create a safe and sustainable water supply as the country's population continues to grow.

ISSUE 2

Damage Caused by Building Dams

Dams store water in order to make water supplies more reliable. However, the building of dams and extra-large dams, known as megadams, can cause damage to natural **ecosystems**.

Dams and Megadams

Dams and megadams store potable water to help create reliable freshwater supplies. For thousands of years, dams have been built to store water and control flooding. Water is run to where it is needed through open channels or closed pipes. Since the 1960s, many megadams have been built. Water in megadams is used for irrigation and **hydroelectricity**. There are now about 50,000 megadams worldwide.

Problems with Dams

Building dams can cause damage to the natural ecosystems of rivers, **estuaries**, and coasts. Fish cannot swim upstream to their **spawning grounds**. Downstream from dams, water flows in rivers and estuaries are often severely reduced. Trees and plants that live along the shores often do not receive enough water. Many dammed rivers no longer flow all year long or flow out to the sea.

Fast Fact

In 2008, water from dams and megadams supplied approximately 20 percent of the world's electricity and 15 percent of the world's irrigation water.

The Three Gorges Dam in China is the world's largest megadam.

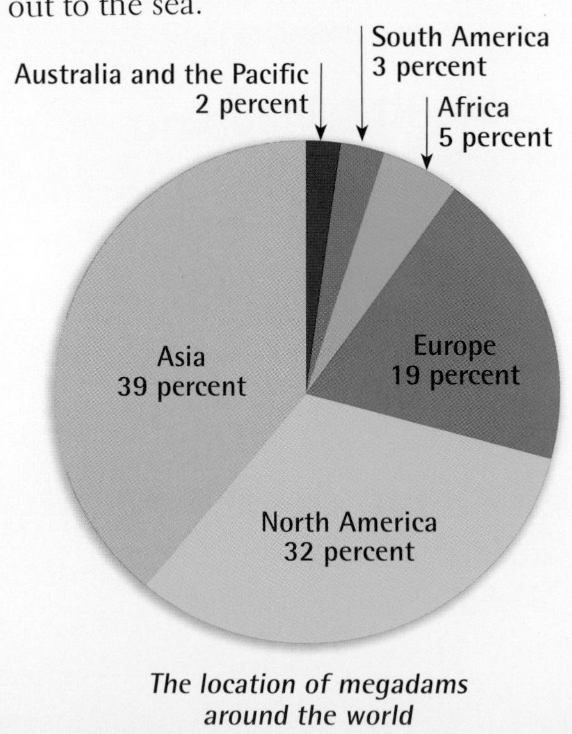

The location of megadams around the world

Nalubaale Power Station takes its name from the Ugandan name for Lake Victoria.

CASE STUDY

Water Levels in Lake Victoria

In 1954, the Nalubaale Power Station was built in Uganda, in Africa, on Lake Victoria. The lake became one of the world's largest water stores for hydroelectricity.

Environmental Impacts

The Nalubaale Power Station and dam produce hydroelectricity for Uganda and parts of Kenya. In 2006, the station reduced its production of hydroelectricity because water levels in Lake Victoria were falling so rapidly. Water levels in Lake Victoria today are much lower than before the Nalubaale Power Station was built.

Cultural impacts

The drop in Lake Victoria's water levels is impacting on a large number of people. About 30 million people rely on Lake Victoria for fishing, transportation, electricity and tourism. It has been suggested that Uganda could build more hydroelectric power stations to reduce the impact on Lake Victoria. However, the suggested sites for new dams are often of cultural importance, such as Bujugali Falls.

Fast Fact

During the 1970s, the building of two or three large dams was started each day somewhere in the world.

ISSUE 2

Toward a Sustainable Future: Alternatives to Dams

Alternatives to dams do not have such large impacts on the environment. Dams can also be designed to minimize their environmental impact.

Alternatives to Dams

There are many alternatives to dams that do not have such negative impacts on the environment.

- Wind and solar energy can be used to replace hydroelectricity, which is generated by water from megadams.
- **Artificial levees** can be used to control floods and protect natural wetlands.
- **Rainwater harvesting** can be used in areas where dams are no longer providing a reliable water supply.

Fast Fact

Rainwater harvesting tanks are one alternative to dams. In the 1980s, more than 10 million rainwater harvesting tanks were constructed in northeast Thailand.

Minimizing the Environmental Impacts

Dams and megadams can be designed to provide for human needs without causing damage to local ecosystems. Many dams today are designed and built to minimize their environmental impacts. Some dams even include migration systems for fish to bypass dam walls.

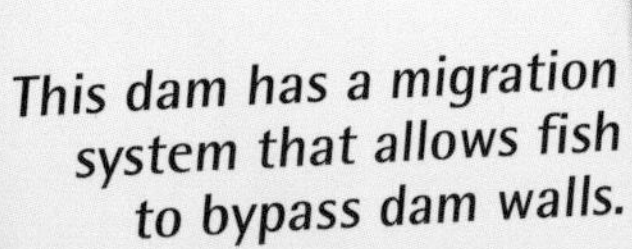

This dam has a migration system that allows fish to bypass dam walls.

Today, the Kennebec River's ecosystem is recovering due to the removal of the Edwards Dam.

CASE STUDY

Dismantling the Edwards Dam

In 1997, the Edwards Dam in Maine became the first dam in the United States to be removed for environmental reasons.

Declining Fish Populations

The building of the Edwards Dam led to a decline in local fish populations. Before the dam was built, the river was rich in Atlantic salmon, striped bass and river herring. However, the dam was built on a tidal area of the Kennebec River and it blocked the river so that these fish could not migrate upstream to spawning grounds. As a result, local fish populations plummeted.

Removing the Edwards Dam

By 2003, the removal of the Edwards Dam was complete and fish populations have since begun to recover. Today, the first 17 miles (28 km) of the Kennebec River flow freely and the river is one of the few remaining spawning grounds for the endangered Atlantic sturgeon. The three megawatts of power that were generated by the Edwards Dam every year could be saved by replacing 75,000 light bulbs with energy-efficient bulbs.

Since the Edwards Dam was removed, approximately 460 other dams in the United States have also been removed. These removals are often completed with the assistance of American Rivers, an organization that helps river ecosystems to recover.

Fast Fact

Today, there are approximately 75,000 dams in the United States. Almost 66,000 of these dams are located on rivers.

Wasting Irrigation Water

ISSUE 3

Nearly 70 percent of Earth's harvested water is used to irrigate crops, but much of this water is wasted by inefficient irrigation methods. In many areas, **salinity** is increasing due to the large amounts of water that are used to irrigate crops.

Fast Fact
Flood irrigation systems can also cause erosion, as soil on the surface of the ground blows away in the wind.

Irrigating Food Crops

Irrigating food crops can help to produce more food, particularly in areas with low rainfall. However, many irrigation methods waste large amounts of water. Flood irrigation systems drench crops in water and much of this water **evaporates** without helping crops to grow. Flood irrigation also increases salinity in the soil. As water sinks into the soil, it makes the **water table** rise and brings salty water to the surface.

Less Water for Irrigation

As the climate warms and rainfall patterns change, there is less water available to harvest for irrigation. At the same time, Earth's population is growing and more crops are being planted for food. Countries with little ground water, such as Egypt, often import food from other countries because they do not have water to irrigate crops. Countries with large amounts of ground water, such as Australia, often harvest **unsustainable** amounts of this water. This can also cause salinity and even **erosion**, as plants die and soil is blown away.

Flood irrigation systems drench crops with water, but most of this water evaporates into the air.

The Aral Sea is now surrounded by salty and infertile land.

CASE STUDY

The Disappearing Aral Sea

The Aral Sea in Central Asia is disappearing as water levels drop and salinity increases. Water to irrigate cotton and rice crops is being harvested from rivers that run into the Aral Sea.

Fast Fact
Almost 5 million acres (2 million ha) of land around the Aral Sea are now infertile due to salinity.

More Water Being Harvested

As the number of crops grown in Central Asia has increased, more water has been harvested for irrigation. Since 600 BCE, people have used water from two major rivers, the Amu Darya and the Syr Darya, to irrigate their cotton and rice crops. Today, the amount of water harvested is so great that no water flows from these rivers into the Aral Sea.

Changes to the Natural Environment

The Aral Sea and surrounding areas have changed dramatically in the past 30 years. The Aral Sea has shrunk to less than one-third of its original size and water levels have fallen by at least 46 feet (14 m). The remaining water has become high in salt and many fish have died. Fishing, which was once an important activity in the Aral Sea, is now impossible.

ISSUE 3

Toward a Sustainable Future: Conserving Irrigation Water

Crops that are irrigated efficiently do not need large amounts of water. Using more efficient irrigation methods and reducing the need for irrigation can help to conserve water supplies for the future.

Using Efficient Irrigation Methods

Farmers can conserve irrigation water by using efficient irrigation methods. Drip irrigation systems drip small amounts of water onto the soil as it is needed. This system requires less water from rivers, lakes, and underground stores. If most farmers used this system, the impacts of water harvesting on many freshwater ecosystems would be reduced.

Reducing the Need for Irrigation

Many farming practices can be adopted to reduce the need for irrigation. Farmers can:

- choose crops that will grow well in their local area, without the need for large amounts of irrigation water
- grow crops of **green manure** between food crops to add moisture to the soil and reduce the need for irrigation
- collect rainwater to irrigate crops by building earth walls to trap water so that it seeps into the soil

Drip irrigation systems are designed to put every drop of water to good use.

Terracing and earth walls can help rain to seep into the soil and irrigate crops.

CASE STUDY

Using Green Water

Rainwater that sinks into the soil and irrigates crops is sometimes called "green water." Green water is the largest single source of freshwater used to grow crops.

Crops that Rely on Green Water

Many crops are watered by green water, including:

- tree plantations
- food crops, such as wheat
- industrial crops, such as cotton

Today, about 60 percent of Earth's staple food crops are watered solely by green water. In some parts of Africa, almost all food crops depend only on green water.

Fast Fact

Mixed cropping, or planting several crops together on the same land, can reduce the need for irrigation. In Morocco, date palms are grown with wheat crops and stone fruit trees are grown with barley crops.

Increasing Soil Moisture

Farmers can increase the amount of rain that is absorbed into soils and becomes green water. Farming techniques that break up soil, such as ploughing, take water out of soils. As a result, soils can be eroded by wind and rain. By contrast, growing green manure crops can help soils to retain water, remain fertile, and produce greater crop yields.

Overuse of Ground Water

People are now using ground water at a much faster rate than it can be replaced. This is creating problems with water supplies in areas that rely on wells and **bores**.

Unsustainable Use

In most parts of the world, unsustainable amounts of freshwater are being taken from the ground. People have long drawn ground water from wells and bores for use in homes and on crops, particularly in dry areas of the world. However, as the population grows and more food crops are planted, larger amounts of ground water are being drawn to the surface.

Fast Fact

Ground water can become polluted by chemicals from industry and farming. In India, high levels of the dangerous chemical arsenic were found in ground water.

Impacts on the Environment

Depleting the supplies of ground water can have many serious consequences for the environment. Rivers, wetlands, and lakes that are fed by ground water may dry out. Seawater may flow into underground stores and aquifers as freshwater is drawn out. This can increase salinity in ground water.

People in villages across the world pump ground water from wells to use for drinking and washing.

The water in the Ogallala aquifer is used to irrigate crops in Texas.

CASE STUDY

Draining the Ogallala Aquifer

The Ogallala aquifer irrigates about one-fifth of crops in the United States. Water levels in the aquifer are dropping rapidly as large amounts of ground water are used for irrigation.

Ground Water in the Ogallala Aquifer

About 95 percent of freshwater in the United States is ground water. The Ogallala aquifer is the largest aquifer in the United States and it stretches from South Dakota to Texas. Ground water from the Ogallala aquifer is being used at unsustainable rates to irrigate crops and farmland.

Unsustainable Harvesting of Water

About 424 billion cubic feet (12 billion cu m) of water are harvested from the Ogallala aquifer each year. This is 18 times the yearly flow of water in the Colorado River. However, this level of water harvesting has become unsustainable. Many farmers have had to change their irrigation methods, such as switching to drip irrigation systems. Some have also had to plant crops that need less water to grow.

ISSUE 4

Toward a Sustainable Future: Using Less Ground Water

Ground water is replaced over a very long period. Ground water supplies can be protected by conserving areas where it collects and using alternative sources of water where possible.

Fast Fact
Ground water can also be supplemented with other sources of water, such as green water and recycled water.

Conserving Recharge Zones

Recharge zones are areas where runoff collects before it runs into rivers and aquifers. Plants and other vegetation in recharge zones prevent water from running off quickly, so that it filters down into the soil. Protecting plants in recharge zones will help to maintain ground water supplies.

Alternatives to Ground Water

There may be other sources of water that can be used as alternatives to ground water. Recycling water is one alternative to drawing more water from the ground. The city of Moscow, Russia, recycles most of its water in enormous water-treatment plants instead of drawing more water from the ground.

Runoff from recharge zones runs into rivers such as these.

It takes millions of years for water to travel through the Great Artesian Basin, which covers about 669,887 square miles (1,735,000 sq km), or one-fifth of Australia.

CASE STUDY

Conserving Ground Water in the Great Artesian Basin

ISSUE 4

Water in Australia's Great Artesian Basin is being conserved by capping, or covering, free-flowing bores and building pipes.

Open Bores and Channels

Thousands of bores in the Great Artesian Basin provide water for irrigating crops in dry areas of Australia. Since 1878, bore water has been tapped and run through open channels. Large amounts of this water evaporates as it moves through the channels.

Fast Fact

Ground water is the main source of water for about two-thirds of the Australian continent.

Capping Bores and Building Pipes

Today, bores are being capped and pipes are being built to conserve water in the Great Artesian Basin. These measures reduce the amount of water that is lost through evaporation. During the 1990s, more than a quarter of Queensland's bores were capped and many pipes were built. Piping water from the Milroy bore at Moree, New South Wales, increased water efficiency dramatically.

ISSUE 5

Access to Freshwater

It is predicted that water shortages will affect more than half of the world's population by 2025. Today, some people are using large amounts of freshwater while others have little access to clean, freshwater.

Water Usage Patterns

Water usage patterns show that some people use much greater amounts of freshwater than others. Earth's population today is approximately triple what it was 100 years ago. Over that time, water consumption has increased twice as fast as the population has increased. On average, people in developed countries use 10 times more water each day than people in developing countries.

Technologies to Increase Freshwater

Many technologies can be used to increase the amount of freshwater that is available for use. These technologies include:

- recycling waste water at water-recycling plants
- removing the salt from seawater at desalination plants
- decreasing evaporation by capping bores and building pipelines

However, most of these technologies are very expensive. Often, only developed countries can afford to build water-treatment plants or pipelines.

Fast Fact
World Water Day is held on 22 March each year to highlight issues related to water supplies. The theme of World Water Day in 2007 was "coping with water scarcity."

Large amounts of water are used to keep this luxury golf course in the Mojave Desert green.

Most of the freshwater that is used in Dubai has been desalinated in water desalination plants such as these.

CASE STUDY

Water Desalination in Dubai

The wealthy country of Dubai uses water desalination plants to supply freshwater to its residents. This water is often used wastefully in swimming pools and gardens.

Fast Fact

Desalination produces most of the freshwater for wealthy Middle-Eastern countries, such as Saudi Arabia and Kuwait.

Water Desalination

Water desalination is a process that removes salt from seawater, so that it becomes safe to drink. It uses expensive technology and only the richest countries in the world can afford it.

Water desalination provides approximately 98 percent of Dubai's freshwater. Dubai is a wealthy country with low levels of rainfall, no rivers, and little usable ground water. Its coasts are lined with gas-fired power plants. These plants turn seawater from the Persian Gulf into freshwater through desalination.

Use of Desalinated Water

Water is often used wastefully in Dubai and other wealthy Middle-Eastern countries. It is used to run fountains, fill artificial lakes and swimming pools, and to water plants at holiday resorts and grass on golf courses.

Toward a Sustainable Future: Sharing Water Supplies

Freshwater supplies can be shared more equally between countries. If the demand for freshwater was reduced, there would be more freshwater available for others.

Fair Access

Governments can make sure that access to freshwater is shared fairly between countries. Countries that share access to a lake or a river can negotiate so that each country receives some freshwater. Fairer access to water supplies can also be achieved by:

- reducing water consumption so that there is more freshwater available
- investing in technologies such as desalination to create new sources of freshwater
- recycling waste water into drinking water

Reducing Water Demand

The demand for water can be reduced. People can use water more efficiently by adopting some simple measures. These include:

- using water-efficient devices, such as low-flow showerheads and dual-flush toilets
- reusing water, such as watering the garden with **gray water**
- harvesting rainwater in water tanks

Fast Fact
Coal-fired power stations, nuclear power plants and hydroelectricity plants all use large amounts of freshwater.

Many people have installed water tanks to harvest rain that falls on their properties.

People around the world march on World Water Day to make others aware of the issues involving water supplies.

CASE STUDY

United Nations' Water Programs

The United Nations runs many programs to ensure that all people have access to freshwater. They also aim to make people aware of the importance of clean drinking water.

Water for Peace Project

The United Nations Educational, Scientific and Cultural Organization (UNESCO) has a project called Water for Peace, which aims to help countries avoid conflict over access to freshwater. It focuses on areas where access to freshwater is shared between countries. These areas include freshwater river basins in the Rhine River and the Aral Sea, and the Limpopo and Incomati rivers, the Mekong River, the Jordan River, the Danube River, and the Columbia River systems.

International Year of Sanitation

The United Nations declared 2008 the International Year of Sanitation. Over two million people in developing countries, most of them children, die every year from diseases associated with a lack of clean drinking water. Good sanitation and hygiene practices, such as keeping rivers free from human waste and other sources of pollution, is essential to keeping drinking water safe.

Fast Fact

In 2006, World Water Day focused on ways of seeing, using, and celebrating water in cultural groups across the world.

What Can You Do? Use Water Wisely

We all need to use water wisely and conserve as much of it as we can. Knowing about water supplies and the demands in your area can help you to understand the importance of using water wisely.

Create a Water Map

You can create a water map for your home or school. This map can show your local freshwater supplies and the demands for water in your area. You can explore the following questions:

Where does your water come from?

- What are the main sources of your water supply? It could be from a river, an aquifer, or a rainwater tank.
- How far does your water travel to get to you? It could be located nearby, or transported from somewhere else.

Who uses water in your home or school?

- Who uses the most water in your home or school? It could be gardeners, cleaners, or your brothers and sisters.
- What do these people use water for? It could be to water gardens, clean, or have showers.

Fast Fact

Organizations such as Waterwatch, Coastwatch, and Frogwatch help to monitor and protect coasts, wetlands, and rivers.

How is water managed in your community?

- Who controls the water sources in your community? It could be the local government or a private company.
- Are any chemicals added to your local water supplies? If so, what are they for?
- Are there limits on the amounts of water people can use? If so, what are the limits?

How does rainfall affect your local water supply?

- Does rainfall contribute to your community's water supply?
- Has the rainfall pattern in your area changed over recent years?

Save Water

You can save water by:

- having shorter showers
- collecting gray water to use on plants and gardens
- using the water-saving option on dishwashers and washing machines
- checking for dripping faucets at home or at school, and asking someone to fix them

Conduct a Water Audit

Follow these steps to conduct a water audit and find out how much water a dripping faucet wastes each day.

1. Record how much water you think will be wasted in one day by a dripping faucet.
2. Place a bucket under a faucet in a bath or sink and turn on the faucet.
3. Turn the faucet off slowly until the water is reduced to a steady drip.
4. After an hour, record how much water is in the bucket.
5. Multiply this number by 24 to see how much water is wasted each day by one dripping faucet.
6. Use the water in the bucket to water plants or wash dishes.

Conduct a water audit to discover how much water is wasted by a dripping faucet.

Toward a Sustainable Future

Well, I hope you now see that if you take up my challenge your world will be a better place. There are many ways to work toward a sustainable future. Imagine it ... a world with:

- decreasing rates of global warming
- protected ecosystems for all living things
- renewable fuel for most forms of transportation
- sustainable city development
- low risks of exposure to toxic substances
- a safe and reliable water supply for all

This is what you can achieve if you work together with my natural systems.

We must work together to live sustainably. That will mean a better environment and a better life for all living things on Earth, now and in the future.

Web Sites

For further information on water supply, visit these websites:

- UN water quiz
  www.un.org/cyberschoolbus/waterquiz/waterquiz4/index.asp
- European Space Agency water world
  www.esa.int/esaKIDSen/Waterworld.html
- UN Water for Life www.un.org/waterforlifedecade/kids.html

Glossary

aqueduct
a pipe or channel that transports water from a natural source to where it is needed

aquifer
an underground store of water between layers of rock

artificial levees
banks that are built up to control flooding

bores
small, deep holes dug to access ground water from beneath Earth's surface

catchments
areas where rainfall runs off into water stores

climate change
changes to the usual weather patterns in an area

consumption
using water

contaminating
making dirty and unsafe through pollution

desalination plant
a water-processing plant where salt is removed from water so that it is safe to drink

developed countries
countries with industrial development, a strong economy, and a high standard of living

developing countries
countries with less developed industry and a lower standard of living

ecosystems
systems of natural connections between living and non-living things in an area

environmental flows
minimum amounts of water that are needed to keep river ecosystems healthy

erosion
gradual wearing away of soil by the wind

estuaries
bodies of water near river mouths, where rivers meet oceans

evaporates
water that changes from a liquid to a gas

global warming
a rise in average temperatures on Earth

gray water
water that has been used around the home for baths, showers, washing, and cleaning

green manure
a cover crop, such as clover, which adds nutrients to the soil and helps to improve and protect it

ground water
supplies of water beneath Earth's surface, created by water that has seeped into the soil

hydroelectricity
electricity produced by flowing water

irrigation
watering land to grow crops

nonrenewable
a resource that is limited in supply and cannot be replaced once it runs out

potable
water that is fit to drink or store to drink in the future

rainwater harvesting
catching and storing rain for later use

runoff
rainfall that does not soak into the soil

salinity
the amount of salt in water or land

sanitation
waste removal and treatment

spawning grounds
areas where fish come every year to breed

temperate zones
areas of land and sea located between the Tropics and the North and South poles

unsustainable
unable to continue for a long period of time

urban
an area with a high human population, such as a large town or a city

water table
the top layer of ground water

Index